Anonymous

Annie Linn, the moorland flower

Anonymous

Annie Linn, the moorland flower

ISBN/EAN: 9783337111588

Printed in Europe, USA, Canada, Australia, Japan

Cover: Foto ©berggeist007 / pixelio.de

More available books at **www.hansebooks.com**

DEDICATED TO MY DEAR NIECE,

ANNIE,

AS A TOKEN OF AFFECTION,

BY THE AUTHOR.

HALIFAX, JANY. 1ST, 1866.

MY DEAR NIECE,

Excuse the liberty which I have taken in dedicating to you these "uncouth rhymes" as a token of affection, which "makes me glad to pay"

> "Such honours to thee as my numbers may;
> Perhaps a frail memorial but sincere."

Not scorn'd by thee, although despised by less indulgent ones. The story is a true one, and I have endeavoured to tell it simply yet faithfully;—do not blame me if it is sorrowful;—there are more cloudy days than sunny ones. Be merciful in thy criticism, though others may not be. But those who anxiously endeavour to show me its many faults will loose their labor, for I know that they abound throughout. But he who can discover its merits deserves more credit for perceptive powers, for they are not so abundant.

> "True ease in writing comes from art not chance."

I should never have presumed to publish it, had it not been for the success which a previous little Poem attained, and the request of several valued friends whom I fear friendship has blinded to my faults. Such as it is I send it to thee as a token of my affection,

> "Tho' poor the offering be,"

and trust that when far away it may call to mind remembrance of

THE AUTHOR.

ANNIE LINN.

I KNEW her—Annie Linn—a pretty child
That in times pass'd has gambol'd round my knee;
And running oft amongst the moorland's wild,
Has vied with moorland flowers, seem'd as free;—
Ev'n now methinks I see her mellow eyes
Of soft ethereal blue, and the gold waves
Of her light Saxon hair, the roseate dyes
Which dwelt upon her cheek, and strove to save
The lily's monopoly of her face,
And make it share possession with the rose;
While each seem'd striving which could each displace,
And each victorious, softly repose
On her sweet countenance; a heavenly smile
Wreath'd round her ruddy mouth, which knew not
      guile.

When Annie Linn was born, near twenty years
Had passed o'er me, yet when she first began
To walk alone, to share her joys and fears
Was pleasure to me : and many a plan
Did I contrive, to keep her near to me,

And her fond parents, kind, good souls, tho' poor,
Would smile as soon as e'er they chanced to see
Me coming near, and open wide the door ;
And Annie, toddling on unstable feet,
Would hurry out to meet me, holding high
Her arms above her head, she would me greet
With lisping accent, call me Uncle John,
Nor would she rest, until safe on my breast
Her head was softly pillow'd ; there were none
Before, and none have since e'er found a rest ;—
Now I'm alone, uncar'd for, and unblest.

And when the summer sun its radiance shed,
And bath'd in brilliancy the hills and dales,
For sweet companionship I often led
Dear Annie forth, and as the sultry gales
Have swept across her little brow so white,
Or wantoned 'midst her curls, and made her gay,
I oft have look'd, and dreaded lest the sight
Of wings should warn me she would fly away ;
For she appeared too pure to be of earth,
Each thing of beauty strove to hide its head,
The heather seemed to pale in early birth,
And the wild dog-rose blushed a deeper red
When she approach'd, for she was passing fair,
A moorland flow'r, as wild, as free as they ;—

Her simple mind knew not the weight of care,
Her silv'ry laugh drove others' cares away,
For melancholy near her could not stay ;—
And thus together did we spend the hours,
Her mind grew with her body's strength, her love
Outgrew them both, and fell in soothing show'rs
On the parch'd heart, as blessings from above.

But time, perpetual mover, as it pass'd,—
Wrought num'rous changes for poor Annie Linn :—
Sickness and death amongst her friends it cast,
And left her without parents, without kin :—
From babyhood to girlhood she had grown,
And nature seem'd as tho' to make amends
For her unkindness, for the joys o'erthrown,
And for thus leaving her devoid of friends,
As tho' she show'rd upon her every good ;
Not dazzling beauty such as makes you quail
She gave her, but such beauty as is woo'd
By painters, who an angel's face unveil,
And fix it upon canvas, there to show
Mortals such beauty as is none below.

And there, when all were gone, beside the door
Of that old cot she lean'd, and wept her grief ;
Tears such as fall from none but from the poor,

The poor with soul too proud to ask relief;
And her long ringlets like dishevelled threads
Of shining gold, clung to her lovely face,
All wet with tears, such tears as sorrow breeds ;
Tears drawn from out the heart's remotest place :
And whilst she stood and wept, I gaz'd and sigh'd,
I could not trespass on her lonely grief,
She wept for those who own'd her as their pride,
For those who own'd her for a time so brief ;—
She left the door ;—gave one long ling'ring look ;
I met her at the gate and took her hand,
With strong emotions her frail frame was shook,
And sobs, half-chok'd, burst forth without command ;—
I knew, poor child, that she could frame no way
To live, nor had she means to gain support ;
Her eyes were turn'd upon me, their dull ray
Told all she felt, although the gaze was short :
She held my hand, and trustingly was led,
Home to my humble dwelling near the wood,
I could not bear the thought, she should be fed
By stranger hands, or that her mind so good,
Should, by contact with vice, be soil'd and marr'd,
Or lose one jot of that angelic beam,
Which sent its glimmer thro' each act and word,
And made each look of heav'nly sweetness seem,
And made me her more heav'nly still to deem.

In time her grief was lessen'd, and I sent
Her to a school, it was not far from home
And 'twas a lovely walk, I often went
To meet her as she'd home at even come ;—
The way was narrow, and on either side
Rose up an earthen footpath crown'd with trees,
Some old, decay'd, and some just in their pride,
And slender young ones bending to the breeze ;
Their branches interlacing overhead,
Rich with their foliage, and closely grown,
Kept a perpetual twilight, or instead
When the bright sun his very brightest shone,
Made rich mosaic shadows on the ground ;
Yet still exclude 1 quite enough of light
That you might feel repose, and not a sound,
Save from some tiny bird of plumage bright,
Broke in upon your peace ; here would I watch,
Stretch'd on the green moss near a mould'ring stile,
And strain my eyes impatiently to catch
Her earliest approach, her welcome smile,
And sometimes hold communion with myself,
And ask my inmost heart if I was true ;
And if I valued not myself and pelf
More than in former years I used to do,
And if I did not value her, and feel
More interest in her than pure friendship's bring,

Or if some tend'rer passion did not steal,
And round my heart its wand'ring tendrils fling ?
And sighing, wait, until with fairy feet
She'd hasten to invade my cool retreat ;

And in the summer evenings, roaming far
To cull wild flow'rs, and breathe the balmy air,
Or watch the rise of the first evening star,
And hear the landrail's croak, as of despair ;
Or sitting under trees 'midst glittering flies
Whose winged brightness court the pale moon's ray,
Or listening to the hum which swells and dies
Of some night beetle passing on his way ;—
And feel that all around is full of life,
Yet all is peaceful, and inclines the heart
To fancy insect hum with music rife,
Like fairy minstrels, each one plays his part ;
And walking homewards through the rustic lanes,
Spotted at intervals with loving pairs,
Who breathe to willing ears their tender pains,
Whilst each the bliss of sympathising shares ;
And on the manly breast in sweet repose
Is laid the head of the all-trusting maid,
And drinking in the honey'd sweet that flows
From out his lips,—nor does she feel dismay'd
Although the twilight thickens, and the way

Is far from home, and one but rarely trod,
Secure in him she dreads no evil day,
And holds him second in her heart to God;
And thus the summer nights stole quickly by,
And winter came and brought its homely joys,
How cheering was the blaze that leaping high
Danced in the grate with roaring crackling noise;—
To sit in easy chairs, with doors secure,
And windows proof against its fiercest blast,
And hear the winds sweep howling o'er the moor,
Each following seeming fiercer than the last :—
To see the table set with china ware,
And the clean cloth as white as driven snow,
And the lamp lit that cast no vulgar glare,
But threw its mild soft beams on all below ;
Whilst sav'ry dishes forced their steams on high,
Each one a gem of culinary art,
And the housekeeper's calculating eye
Saw each one placed in its assignèd part,
And when the meal was o'er, smoking my pipe,
And pict'ring objects in the soothing cloud,
Rising in graceful curves, a fitting type
Of by-gone pleasures, such as daily crowd
Upon my brain, for mem'ry kindly lends
Her aid to view again those happy scenes,
Again I hold communion with lost friends,

And feel joy through her retrospective gleams :--
Sometimes would Annie read to me a tale
Of touching pathos, not as school girls read,
Wearily dragging on in sleepy wail,
But tune her voice to suit the written deed,
And I should sit and listen quite entranced,
Happier by far than had I own'd a throne,
Each sentence by some fitting act enhanced,
Each circumstance told in befitting tone ;
And when the hero wept, I wept as well,
And when he laughed, I laughed to help his glee,
And when he had mischance my spirits fell,
And when he claim'd, he found my sympathy ;—
Then Annie, when the hours were growing late,
Would read a chapter from the Holy Book,
And pray for all in high or low estate
That by His grace they ne'er should be forsook,
Then with a blessing she would go to bed
And leave me pondering, when oft alone
Have I in humble prayer bow'd down my head,
And asked God's blessing for the orphan one.

And well do I remember now the morn,
That brought a message seal'd for Annie Linn,
The postman loudly blew the twanging horn
That signal'd his approach with welcome din ;—

The rain fell down in torrents, and the wind
Drove it furiously against the panes,
And seem'd a voice of warning to mankind,
That Winter comes and Summer's glory wanes ;—
And when he placed the letter in my hand,
And on the superscription fell my eye,
There came a dread my heart could not withstand,
A fear that some calamity was nigh ;
And thoughts portending ill rose fast and thick,
With heavy heart I bore the missive in,
My mind convinced it was some cruel trick
To steal my moorland flower, my Annie Linn ;—
With eager eyes I watch'd her break the seal ;—
In mute astonishment she read it through,
And then the old sweet smile began to steal
Upon her features, and her eyes so blue
Look'd full into my face, and then she said—
" 'Tis strange, I thought I was the only one
That now was left,—and that all others dead
Without you in the world left quite alone ;
But here I read that one is on his way,
More lonely and more desolate than me,
Intending but a passing call to pay,
To learn the stories of their deaths from me :—
He is a soldier, and his service done,
He comes back to his native land to dwell,

And finds save one, each kindred tie is gone,
Me only left, the mournful news to tell.
If I mistake not, I have heard you say
That you remember when he went abroad,
A thoughtless youth with lightsome heart and gay,
He was a loss his friends could ill afford ;
And that his parents died with broken hearts,
Because the news was brought that he was dead,
And now he comes again from distant parts
To mourn for those who all their tears have shed ;—
To-morrow is the day he does intend
To call upon us, so we must prepare
And give him cordial welcome as a friend,
And yet one other favor I would dare
To ask, that he his visit may prolong,
That we may try to cheer his aching breast,
So for a time to win him from the throng
And bustle of the world to be at rest."

" Nay, Annie, a more cheering welcome still,
More than a friendly greeting he can claim,
He is your cousin, so with hearty will
Prepare to welcome,—pray thee do not shame
My hospitality,—see all is done
That these small means afford, so he may see
He is not in the world so much alone,

Nor wanting friends, nor wanting sympathy ;
And beg him he will make my home, his home,
Until the dreary Winter shall be pass'd,
Until the welcome Spring again shall come
To sooth with zephyrs those who've borne the blast:—
I leave thee now sweet ' Hebe' to prepare
The fitting welcome to this son of ' Mars,'
Whilst I, as ' Fama,' go perform my share
To spread the news of who comes from the wars ?
For many of the villagers can tell
Of his departure and the false report
Which told his death, and on his parents fell
A heavy blow which cut their sorrows short."

'Tis strange how much of selfishness can dwell
Within the heart of man !—what envy sits
Enthroned within the breast !—nor can he tell
From whence it comes ;—until some object flits
Across his path, he lives in fond security,
And deems himself most saintly, and ignores
The truth, that ought without the garb of purity
Could gain admittance at his bosom's doors ;
And oft with smooth complaisancy will dream
That he is safe,—his righteous armour on,—
When as a stone drop'd on a tranquil stream

B .

Some trouble falls,—and lo! his peace is gone.—
" *Man know thyself!* " —'Tis better he should not,
Lest finding out the multitudinous ills
That constitute himself,—what passions hot
And ready for revolt his bosom fills
He should despairing,—give him up for lost,—
Nor think that justice could be still withheld
By mercy,—for incessantly being cross'd
E'en mercy's intercession be repelled.

Thou, love, which holds a prisoner my heart!
Ye tempting, whispering hopes, begone ! begone !—
Base cowardice, usurp each vital part
And make me tremble, when I look upon
Her peerless loveliness ;—lest driven bold
By fear of loosing the ecstatic bliss
Of drinking in such honey'd words as roll'd
From out those lips,—I dare to do amiss,
And tell her that, I love her !—strive to win
Her heart, and change the current of her love
From out its even course,—so Annie Linn,
May, taking her adopted father, prove
Her gratitude,—Hence ! hence ! avaunt base thought,
Much rather would I, yond heav'n kissing hills
Were hurl'd upon my head, that act so fraught
With dangerous consequences that it fills

My soul with fears, should emanate from me !—
Ye evil thoughts ! which like rank parasites
Feed on my better self, nor leave me free
One moment, but intrude on my delights,
Which wanting you were pure as dews from Heav'n,
How came ye thus to tempt me !—cease your wiles !—
For rather should my throbbing heart be riv'n
By ceaseless anguish ;—than one of her smiles
Should be a ray less bright,—and me the cause,
For purest light, the deepest shadow throws.

At last ! the hour approached when he should come,
In dreading restlessness I paced around,
As though I waited for some fearful doom ;—
I listened, and I fancied every sound
Seem'd tuned to some, sad, melancholy note,
The sighing winds,—the rustle of dry leaves,
The sparrow's chirp, who for his breakfast sought,
All sounded dreary ;—quickly fancy weaves
Dark thoughts of some approaching evil hour
And makes the strong man coward,—and his heart
Sinks heavy in his breast ;—then the power
Of fear and hopelessness each bear their part,
And hold him down in their tyrannic chain,
Till Faith, sweet Goddess, sets him free again.—

Why should I dread young Aplin's visit thus?
A youth come from a foreign land alone,
Where he had singly fought the sturdy Russ,
Or held his part in battles nobly won.—
Was it not right that he should seek to trace
The only tie of kindred left to tell,
How death had claim'd each member of his race
Whom when he left were all alive and well ?—
Yet still I felt a fear, lest he should see,
And seeing love the beauteous moorland maid,
And leave but her remembrance to me
" To cheer me as I walked life's checquered shade."

The sound of hoofs ring now upon the road,
'Tis he ! he reins his charger at the gate,
And leaping lightly from his saddle broad
Strides to the door with measured martial gait ;—
His dangling sword clanking against the ground,
His glitt'ring helmet with its nodding crest,
His manly face by long exposure brown'd,
And silver honors thick upon his breast,
His figure cast in true Herculean mould,
His lips compressed, show'd his determin'd mind,
And heavy eyebrows fringed his forehead bold,
A careless smile making the whole seem kind :—
And knocking with his whip, as though the door

Barred up the entrance to some castle keep,
Starting the old housekeeper with the roar
Till she could scarce command her aged feet.—
A noble brute of pure Newfoundland breed
Stood at his feet, and peer'd into his face,
Its panting sides told of the headlong speed
Which had been used to gain a resting place ;
The door was opened and I hastened out,
With strange misgivings, lest my face should show
He was not wholly welcome, cause a doubt
To cross his mind whether he was or no.

I loathe that man, who whilst dissembling,
Can clothe his face with seeming honesty,
And smiles whilst in his heart assembling
Dark plots, to loose, at opportunity ;—
And breathes kind wishes, gives his open hand
As token of sincerity and love ;—
Knowing !—I'd rather grasp a burning brand
With naked palm,—than touch him with my glove.—
And yet I find myself without a cause
Nursing distrust, which like rank weeds unchecked,
Thriving apace, each virtuous thought o'erthrows,
And lives and feeds on their sweet blossoms wreck'd.

He grasp'd my hand, and gave it hearty shake,
Nor wanting invitation, strode about
In soldier fashion, as though bent to make
Himself at home :—nor was the dog without
A spice of that non-chalance which seems bred
In those who mix up in the world's rough throng,
For in full confidence it laid its head
On the housekeeper's lap, nor deemed it wrong,
To stretch its weary length upon the floor
And with closed eyes in conscious safety snore.—

Now, Annie, like a flutt'ring dove approached,
Clad in sweet modesty, which served to deck
Her lovely face in blushes, that encroached
And bathed in rosy tints her swan-like neck;
And with extended hand and smiling face,
And those blue eyes, which like translucent lakes,
Seem'd full to overflow at his sad case,
Whilst he those fairy fingers doubting takes ;—
And then, the brave and dauntless stands dismayed,
Abashed ! embarrassed ! in confusion thrown,
Feasting his eyes upon the beauteous maid,
Mute as a statue,—rigid as a stone ;—
Recovering himself he bow'd his head,—
His tongue unloosed,—his thoughts' interpreter,—
"'Juno' had ne'er been queen of heav'n," he said,

"Had Annie Linn been seen by 'Jupiter.'—
Forgive this freedom fair one, I am prone
To speak the thoughts that flit across my brain,
But willing for my weakness to atone
If hastily I cause another pain.—
Here have I come to hear the woeful tales
Of long lost friends,—their absence to deplore,
But that sweet face o'er all my grief prevails,
And lights my heart, which thought to joy no more;—
I scarce dare ask a heav'nly nymph like thee
To tell me that sad news,—lest I forget,
And hear the story of their misery
Without a pang of sorrow or regret.—
Come, Nero! faithful brute, we must not stay
And cast a shadow o'er so bright a star,
But hence and bear our darkening gloom away,
And seek for solace in the moil of war."

" Dear Sir, I said, pray be not resolute,
Leave not so suddenly, and bear the gloom
Upon your face, like to a sable suit
To tell the world your thoughts are in the tomb ;—
Come, Annie, with your ever winning tongue
And press the offer, so we may prevail
On his acceptance, for the will is strong,
If thy solicitations too shall fail."

"Cousin, pray thee do not leave me,
Other kindred I have none,
Let me of thy griefs relieve thee,
Till each gloomy thought be gone ;—
And if smiles can give thee pleasure
Gladly will I strive to smile,
Truly happy beyond measure
If thy hours I can beguile ;
See yond fleecy clouds reposing
In the murky vault above,
Each in ceaseless change disclosing
Beauties, as they onward rove,
Short time since in heaviest gloom,
Gilded now in silv'ry light,
As the sun's bright rays illume
The dark curtains of the night;
So let me thy griefs dispel,
Chase the darkening thoughts of woe,
Leaving nought but joy to dwell
In thy breast, then leave us so,—
Share with me my guardian's heart,
There is still some room to spare,
Consolation to impart,
Naught but goodness dwelleth there."

"Cease your entreaties for I fain will stay,
And share with you this peaceful home awhile,
And hear no more the brazen bugle bray,
Which for long years has call'd me out to toil;
But midst fair nature's charms, drink from her cup
Of consolation freely filled for all,
With sweets of comfort, brimming brightly up,
Unmix'd with worldly dregs, of bitt'rest gall."

Thus he consented and became my guest,
And joined in every innocent pursuit,
Which was our wont to do, with hearty zest;
And like his master, the sagacious brute
Shook off his former self, and frisked in play,
Devoid of cares, light-hearted, free, and gay.

And often we have saunter'd out at morn,
When the hoar frost his glitt'ring robe has flung
O'er all the landscape, and the echoing horn
Of distant huntsmen in the hills has rung;
And watched the day's red orb with fiery train,
Struggle to pierce the mists with cheering ray,
Tinting with rosy hue the whitened plain,
Bathing in golden light each frost hung spray;—
And having gain'd the pathway through the wood,
Have wandered on beneath the avenue

Of trees, which like huge marble columns stood ;
And sunlight pierced their close knit branches through
As though we walked in some cathedral,
Where "storied windows" dyed the soften'd beam
To paint in fleeting brightness on the wall
And pierce the gloom with many color'd gleam :—
And here and there a solitary bird
Sat chirping in disconsolate distress,
Or some faint-hearted conie that had fear'd
We should invade his solitaryness,
Plunged in the thicket, which when Nero spied,—
Dashing along he gave it bootless chase,
Returning panting to its master's side
With downcast looks as though in sore disgrace.

Aplin and Annie often lagged behind,
Whilst he with eloquence described the things
Of beauty, or they listened to the wind,
That sighing soothes " Eolus " as it sings ;—
And as she stood with health's glow on her cheek
Beneath those hoary trees, one might mistake
Her for bright " Salus," who had come to seek
" Silvanus " in his bowers, and bid him wake ;—
Or stooping down o'er some small ice-bound pool,
Conning the frost-wrought tracery that mocks
The artist's skill, and proves how weak a fool

Is he who would outvie dame Nature's works :—
And when their eyes have met, a long, long pause
Has intervened, and not a word has pass'd,
And each embarrassed with no other cause
Than that short glance,—their eyes around have cast
To note some object to renew again,
The seeming trifling yet absorbing strain.

Those who have ever felt the power of love
Need no description of its many signs,
The simplest actions which undoubted prove
The wounded heart, and press its sure designs :—
The look of joy, the twinkling of the eye,
The silent tongue, and absent mind, convey
A world of meaning to the lover nigh,
And tells him there is hope, although she may
Avoid him, and in dark frowns veil her face,
He reads her heart as though a crystal vase.

I needed no one versed in " Cupid's " laws
To solve the mystery, the reason why
She often blushed without a seeming cause,
Or heaved, to her almost unknown, a sigh :—
Oh ! how I struggled with my sense of right,
And argued that I had no cause to fear,
My heart was proof and gloried in its might,

No wrongful thought could gain an entrance there,
Strong in my good resolves, their powers to dare.

————

That man is weakest, who sure in his strength,
Places no sentinel upon his heart
To guard 'gainst evil's entrance, till at length
The foe approaching at his weakest part,
 Do so assail with force well disciplined,
They make the breach, and e'er he is awake
The conqueror's banner floating in the wind
Proclaims the spoilers triumph,—his mistake,—
For palsied with defeat he wants the power
To cope with those whose strength he scorned before.

————

And now what most I fear'd had come to pass,
For he had seen, and lov'd, and not in vain ;—
I knew she did not love myself the less,
Her heart was far too pure for such a stain :—
But when at home, sat round the humble board,
And sweet "Felicitas" her influence shed
Upon the scene, it would but ill accord
Had I the cloud of gloom about them spread,
And as he told the tales of war and toil,
And recollection brought the hot blood back
Into his cheeks, fading away the smile

Which left determination in its track ;—
Her mild blue eyes with an unwonted fire
Would sparkle, as the tale in interest ran,
And show the strong, tho' unexpressed desire,
And wish " that Heaven had made her such a man."—
He had not made avowal of his love
Nor trusted his heart's secret to his tongue,
His every act abundantly could prove
Without the aid of words, the passion strong
Which held him captive, feeble as a child,
And knew no sunshine, save when Annie smiled.

I will not dwell upon each stolen glance,
Nor chronicle each sigh or kindly word,
Or note the smile upon the features dance
When the expected footseps they have heard ;
My pen is wielded by a hand too weak
To render justice to such lofty strain,
And leaves the reader in his heart to seek
For light, to make the forced omission plain.

But Annie's busy fingers still would ply
The needle, for she never once forgot
Or left undone the work of charity,
Which made her blest in every humble cot ;
She visited the poor, oppressed with age,

And watched beside the bed when death was near,
Whilst Aplin read aloud the sacred page
And breathed a prayer in accents most sincere.

At last the great momentous day arrived
When Annie, falt'ring asked for my consent
To give her hand to him, and though she strived
To look composed the color came and went
Like cloud-cast shadows o'er her anxious face ; —
She ne'er had asked for aught and been denied,
I could not say her nay in such a case :—
I kissed her brow, and drew her to my side,
And wished her joy as Aplin's future bride.

" But, Annie dear, think on my poverty,
Rich in another's love thou needs not fear,
But there is none to take the place of thee,
Robbed of the prize my heart has held most dear ;—
No man is poor because he lacks the joys
Which others have, and which he never knew,
But he, whom fated one of fortunes toys,
Blessed with abundance, or denied the few,—
Who bears caress or slight alternately,
Endures more want for luxuries once known
Than he, who born in stern necessity
Can comprehend ;—and thus left so alone,

Unpitied, struggling, oftentimes despised,
To brood o'er former joys too dearly prized."

"Say not you will be poor, I still shall be
The same to you, for naught can e'er dispel
Your image from my heart :—such poverty
You need not dread, for here I still will dwell
Though poorer far than I know words to tell.
What is it to be poor? the want of gold?
For which poor mortals oft risk life and soul?
To bear a part in misery's untold?
Or go a houseless one from pole to pole?
Or is it to be poor when on the bed
Of want and sickness, no fond friendship's nigh,
No loved one near to soothe the troubled head,
Nor one to heave the sympathetic sigh?
Yes! such is to be poor, but poorer still
Are they, who blessed with kind and gen'rous friends
Receive their kindness, and not wanting will
Have not the power to make them just amends ;—
To be so poor in words, that thanks sincere,
Die e'er they pass the threshold of the heart ;—
Such is my poverty, but do not fear,
Ingratitude shall never form a part
Of her, on whom your kindness every day
Has placed a debt 'twill serve a life to pay."

Time hurries on, how quick each moment speeds,—
How swiftly fly the hours when "Saturn" calls
"Aurora" forth to lash her fiery steeds
And wake a day of joy ; and as she rolls
Along the heav'ns in her bespangled car,
Flinging the glitt'ring brightness from the wheels
In silver show'rs, until the waning star
Its nightly lamps soft ray at last conceals ;—
One oft has felt desire to stay her flight,
Regretful such a day should end in night.—
  But when dull melancholy hangs abroad
Her tear-stained ensign, and envelopes all
In rayless gloom, how weary seems the road
Through one short day, when sorrows clouds appal.
  Years have seem'd months, months weeks, weeks
        scarce a day,
Days hours, and hours have pass'd unheeded by,
When void of care along the flow'ry way
Of peaceful life, with sweet felicity
Smiling at every step to cheer us on,
With harmless pleasures ; such like happy days
Were they, but now, alas, those days have gone,
Leaving their wreck round which fond mem'ry plays,
Striving to deck the past with living leaves
Of present thoughts, which wayward fancy weaves.

What more for Annie's good could I desire,
A noble looking youth, whose mind well stored
With ancient lore, one whose thoughts sought higher
Than God's revolving footstool, for he soared
And seem'd most happy in his thoughts of Heaven ;
Richer by far in hopes of future bliss,
Seeming to live to claim the promise given
Of endless joy, and more content with this
And an approving conscience, than the praise
Of selfish worldlings, or the gold reward
Which blinds one half mankind, until they chase
The phantom wealth, and treat with disregard
Those mines, in which more lasting riches hide
Than dwell in gold or in loose tongues abide.

A shallow pated fool with well fill'd purse
May gain applause, from those more senseless still,
Whilst genius dwells despised beneath the curse
Of poverty, and cringes to the will
Of those who own the glitt'ring sordid dross,
And yet disgrace God-like humanity ;—
Still toiling on beneath his weighty cross
He bears their sneers with forced urbanity,
And bows the knee to gild insanity.
    But who, of all who ever felt the joys
Of self communion, and from the mind

C

Has dragged the pleasing thought which never cloys,
And lived in fancy's realms; leaving behind
All worldly cares and troubles, revelling
On fond imaginations soaring wing,
Would change those thoughts for all the drivelling
Of brainless mortal though he were a King.
Or loose for gold, or other earthly prize,
The privilege of living with the dead,
And mingling with the spirits in the skies
Of those illustrious great ones that have fled,
Yet left behind a halo round each name,
Which ages waft into a blaze of fame.

The day was fixed, and with sure steps advanced,
When he should claim the promised hand, and take
Unto himself the priceless gem enhanced
With all the comely virtues which could make
A woman lov'd; and preparations buzz
Early and late, told of the great event
Which fill'd the neighbouring matrons with such fuss
As in and out they ceaseless came and went;—
But Aplin shared not this, for he was call'd
And had to leave in haste one early morn,
To settle weighty matters we are told,
But left assurance of his quick return.

The village bells peal'd forth in gladsome strain
And woke the echoes of the neighbouring hills,
Floating their joyful tidings o'er the plains,
Joining their cadence with the murm'ring rills,
The village was astir, and streamers gay
Waved o'er many a neat tho' humble cot,
And festoon'd flowers were hung to deck the way,
And banners trimmed with the "forget me not;"
Whilst little children dress'd in snowy white
Each with a small bouquet upon her breast
Hurried from spot to spot, their chief delight
The novelty of being gaily dress'd ;—
And stately dames with stiff portentious skirts,
And formidable caps with double frill :—
And sturdy men, sporting their ruffled shirts,
Uncomfortably grand, were standing still
And smiling as they saw each youthful group
Careering past, big with some new event,
Or watch'd the idlers in straggling troop,
Wind up the street, each on enjoyment bent.
Around the church was stood a motley crowd
Who at the beadle laugh'd and poked their wit,
His pursed up mouth and knitted forehead show'd
How oft their playful sallies made a hit,
And he revengeful for his honors slight,
Fiercely enraged, walk'd with more pompous gait,

Scorning to think that gold and scarlet bright
Should fail to strike with awe each rustic pate ; —
And on the hapless urchins who had dared
To venture on the ground he stood to guard,
His sounding whacks impartially he shared
And made them fly the consecrated yard.

    Whilst all abroad seem'd joy and sweet content,
I sat and mused nursing my bitter pain ;
Or softly stealing pass'd the room I went
To gaze on her I ne'er might see again ;
And there she sat, more lovely than before,
Deck'd in her bridal robes, her angel face
Methought a touch of melancholy wore
Which seem'd to add to her another grace ;
She saw me pass, and in a falt'ring tone
Pronounced my name, I hurried to her side,
And found her weeping in her room alone,
Tearful, yet joyful as became a bride ;—
And round my neck she twined her jewell'd arms,
Sobbing she knew not why upon my breast,
Unconscious or unmindful of her charms,
Feeling in happiness prospective blest.

    "Oh think on me, dear friend, when I am gone,
More than a father you have been to me,
I grieve to leave you here again alone,

Dearer to me than all, save only he ;—
He who now comes to claim me for his own,
And oh, my heart shall never once forget
The thousand kindly acts towards me shown,
Until my day of life in night be set ;—
And then, if the blest privilege be giv'n
To mortals when they leave this earthly clay,
To bear the sweet remembrance up to Heaven,
'Twill serve to cheer me in that endless day.
To utter thanks I cannot frame my tongue
As I could wish, but my fond flutt'ring heart
Swells with its gratitude, nor is it wrong
At such a time to sigh that we must part.
And when far distant at the time of prayer,
Ever remembered will I pray for thee,
And oh deny me not a little share
In all your thoughts, your joy, your misery,
I hear your heart's loud beating, now as though
Some hidden sorrow struggled for release,
Speak a few words of comfort e'er I go
And bless me, then can I depart in peace."

" What words of comfort would'st thou have me speak,
'Tis rather me who need solacing now,
Light-hearted smiles will dance upon thy cheek
And hope with joy contend upon thy brow,

And if there lurks a shadow of regret
Within thy heart, the light of love will scare
The gloom away, but oh! my soul would fret
If in thy love I ceased to have a share.
　　Go with my blessing, dearly loved one go,
Thy light should not be hid, the virtues rare
Which live within thee pure as falling snow,
Thy tender heart, which never knew a care
Save those which sympathy implanted there,
Will throw a lustre through a wider sphere
And more shall learn thy pow'rs to soothe and cheer.
　　One short hour more and I shall give away
The key which holds my joys in custody,
Joying yet sorrowing on thy marriage day
And bowing to the loss the fates decree;
God bless thee,—bless thee,—fair creation's gem,
Unrivalled, brightest in her diadem."

"Oh say not pure, I know my froward heart,
And see dark spots which need repentant tears
To wash away, and sinful thoughts that dart
Across my mind, which none save conscience hears,
But let me haste away lest Aplin chide,
Or deem his Annie a reluctant bride."
Then quickly borne along, our anxious eyes
Soon saw the time-worn village church arise.

Thou venerable pile! never forgot
Will thy old features be! like some dear face
Thy form arises clear;—cares cannot blot
Thy image out, nor aught save death erase
The thousand recollections which entwine
Around thy mem'ry: as the ivy grasps
Thy crumbling tower, and woos the fickle wind
To whisper in its leaves, so fondly clasps
The old associations around me,
And binds me to thy sacred shades with ties
Which hold me firmer as my moments flee,
Nor e'er will loosen till remembrance dies.

I seem as though I saw thy walls to day,
Dappled with sunlight, and the sculpture strange
And uncouth, basking in the cheering ray
Of golden sunshine,—which lit up the range
Of weighty buttresses and peep'd between
The heavy mullions of the window'd nitche,
Where formerly the pictur'd pane had been,
To clothe th' intrusive beam with colors rich
As tho' to fit it with more comely grace
And make it worthier of that holier place.
And that old tow'r with battlemented top,
Has half its walls concealed with living green :—
The old sun-dial which had long forgot
To note the hours, weary of what had been

Is now no more than a projecting block
Of shelving stone, unlettered and o'ergrown
With moss and litchen, whilst below, a clock
Has made an ancient window seat its own.
The sheltering porch whose crazy roof was laid
On undressed beams of sturdy English oak,
And proved on stormy days a welcome shade
For outdoor gossips, or the needy folk
Who houseless wandered by; and the one bell
Which for long years above the gable hung
In sullen silence, but old men could tell
Of times of dread, when that old bell has rung
In the mysterious hours of night
When mortal was not near, and has foretold
Some dire calamity; and still with fright
They tell of deeds its knell served to unfold.
　　I led dear Annie up that solemn aisle,
And in the " dim religious light " she seem'd
Like some fair wand'rer in that ancient pile,
And the sweet smile which o'er her features beam'd
Was answered by a hundred smiles around,
And fervent blessings and good wishes fell
From every lip, and thick upon the ground
Where " Flora's " fav'rites culled from hill and dell,
Which scarce seem'd bruised beneath her dainty tread
And all around their fragrant odours shed.

But Aplin had not come, why this delay?
Did he mistake the hour he should appear?
Some unforeseen event might cause his stay,
Some accident, else he had sure been here :
Or sickness perhaps may hold him back awhile,
Or circumstance o'er which he lacks control :—
Why fades from every face the kindly smile?
List! 'tis the bell's prophetic warning toll :—
And Annie heard it and her color fled,
And anxious looks o'erspread each pallid face,
As though some apparition from the dead
Had howling paused, within that sacred place ;
And Annie's frail built form shook like a leaf
And paler grew her cheeks, her eyes were turned
Heavenward, and in whisp'ring accents brief
Pray'd God that Aplin might be safe returned.
Then fainting in my arms I bore her thence,
A poor pale flow'r of blighted innocence.

All sounds of joy were hushed ; no funeral
Had ever cast so deep a gloom around ;
A panic seem'd to seize the hearts of all
Who heard the bell's unearthly ringing sound,
I bore her home, but sympathy was vain,
She lived in death, and never smiled again.

How anxiously I counted every hour
In hopes he would return again, but no !
Weeks rolled away, until at length the power
Of hope was gone, and naught was left but woe.
    At last the curtain fell ;—the mystery,
Was mystery no more, but clear as day :—
A man acquainted with his history
Was thrown by accident across my way,
He told me Aplin was not what he seem'd,
That his religion was most base deceit,
A reckless libertine, who never deem'd
To trample youth and virtue 'neath his feet
A crime ; and then to make his tale believed
He showed me love tokens, the sight of which
Told but too plainly how we were deceived,
And my heart sank pow'rless into the ditch
Of dark despair ; to see each little toy
Which Annie had selected with such care,
O'er which she'd pondered in her childish joy
And falsely dream'd he would her pleasure share.
I bought them from him for a trifling sum,
I have them yet, they tell a mournful tale ;—
Dear fond memorials tho' ye are dumb,
Food for reflection when all others fail.
    Still, still, I hoped more cheering tales to hear,
Alas, in vain, thicker and blacker grew

The cloud of wickedness, and doubt and fear
O'er his long absence dark misgivings threw.

And Aplin came no more, no he had fled
And left behind him a poor breaking heart,
Himself to every voice of conscience dead
To play amongst the world his evil part :—
  Oh why was he not well content before ?
Why envied he the lovely moorland flower ?
Why should he rob what he could ne'er restore ?
Why cheat the maiden of her heavenly dower ?
Perchance e'en then in some low brothel laid,
Half mad with wine which never knew the grape,
Gazing with leaden eyes on some lost maid
And fancying beauty in a harlot's shape ;
Or pacing round the tawdry gas lit room
Which like its inmates dreads the light of day,
Courting with heedless haste his awful doom
And revelling in lust, yet seeming gay,
Whilst the loud laugh which leaves no joy behind
Breaks in upon the peaceful hours of night,
Or round his neck a maiden's arms entwined,
Which clasp another's with as much delight :—
  Or perhaps with one more bawdy still than they
Holds favor'd converse,— opes to her his purse,—
Her who in that her little hell holds sway

And issues forth her orders with a curse.
    And this is fancied happiness to some,
Degraded men !— but human beasts at best,
Devouring innocence where e'er they roam
And slighted virtue furnishing their jest ;—
    Ye heartless hypocrites, with fawning smile,
Wand'ring like rav'nous wolves amongst the sheep,
Using religion's cloak to hide the guile,
Think not that its disguise will aways keep,
But fallen off disclose your hideous form
Bare to the world, fit object for its scorn.

A fading lily laid upon the ground,
Pluck'd by some ruthless hand, then cast aside,
Sweet in its wreck it throws its influence round,
Its fault was beauty, and for that it died :—
Sweet flower, fit emblem of poor Annie's lot,
Sought for, obtained, discarded, and forgot.

The pious priest with calm impressive face,
In fancy I appear to see him now,
Bearing from cot to cot the means of grace,
Braving the pelting rain or blinding snow,
His broad-brimmed hat, 'neath which his hoary hair
Hung straggling from his venerable head,

The quaint cut overcoat he used to wear,
Figure erect, his slow and stately tread,
His oaken staff 'fore whose imposing shake
The village urchins ran with well feigned fear,
Whilst sternly frowning still he fail'd to make
Them to regard him as a friend less dear.
And he as Annie pined away would come
And strive to cheer her, bid her hope again ;—
Her drooping eye and heaving breast tho' dumb
Told more than words her life's devouring pain :—
And day by day as she more sickly grew,
Paler and paler, sinking like a child,
Her eyelids shutting out the world from view,
Her face clothed in the old expression mild :—
Whilst he in prayer bow'd down his reverent head
Beseeching God to spare her for awhile.
"Thy will be done, not mine, oh God," she said,
"Thy chast'nings I can welcome with a smile."
And many times the minister would stay
And brush the tear away that dim'd his eyes,
With falt'ring voice he'd strive to teach the way
"To realms of bliss prepared beyond the skies."
And she was lost to all her cares below,
Her thoughts seem'd all to yearn to be above,
Content to leave this world of pain and woe
And firmly trusting in a Saviour's love.

And then the bustling doctor, cane in hand,
With look mysterious and talk profound,
Did visit daily and give some command
And leave in haste to go his 'custom'd round.
   But once he came, nor left he her again,
But anxious stood and watch'd her flick'ring breath
Exerting all his craft to ease her pain
And bravely striving to out master death :—
   At length his eyes began to sparkle bright,
And bustling women hurried here and there,
Each seeming to partake of his delight
And each in the good fortune claim a share.
At last he placed upon the nurse's arms
A little smiling cherub newly born,
Sweet innocent, with all its mother's charms,
Doom'd not to live to spend life's fleeting morn.
   But Annie now demanded all his care,
Windows close curtain'd to exclude the light,—
Each face look'd anxious, yet not one would dare
To ask the other if he guessed aright ;
And soon she slowly raised her drooping eyes
And look'd around on each familiar face,
And tries to speak, alas, she vainly tries,—
Then sank exhausted with her hopeless case,
And there she lay whilst stimulating drinks
Revived once more the last small vital spark,

Which like the waning taper e'er it sinks
Shines brightest, and is then for ever dark.
   She summoned me, I hastily complied
To hear the words I knew would be her last,
"Tell Aplin when you see him that I died
And blessed him, and forgave him for the past."
   "Grant me your blessing, let me lay my head
And part with life upon your faithful breast,
Death and the grave have lost their powers to dread,
Good bye, she sighed, and calmly sank to rest."

   Scorn not the fallen, tho' far, far astray
Their feet have gone, oft trusting innocence
And unsuspecting hearts are forced to pay
The penalty of keen remorse ;—the sense
Of conscious degradation will depress
The heart for sins which are not all their own :—
And wrongs enduring, for which no redress
Can be obtained, or kindliness atone.
   Let those not boast of virtue, who ne'er stood
Within the shade of strong temptation's power ;—
Nor talk of chastity, who 'gainst the flood
Of life's deceptions have shut fast the door,
Or hid behind a convent's walls the light
God gave them, as a ray to cheer and guide

Some one less fitted for the arduous fight,
Who wanting help has thoughtless step'd aside.
  But they who bravely battle with the world,
And through a lifetime press on for the goal
Of duty done, their tattered flag unfurl'd
Merits more praises, tho' at last they fall.
  The soldier who at home reads o'er the tales
Of hard fought battles which he never shared,
May charge with cowardice the force that fails
And flees before the foe, or that which dared
To grapple to the last, and conquered lies,
He may condemn for want of judgment shown,
Whilst pulseless 'neath inhospitable skies
Lie braver hearts than his, cold and unknown.
  Despise not then the frail one sorely press'd,
Whose fault was innocence—but let her rest.
" Admitting her weakness, her evil behaviour,"
" But leaving with meekness her sins to her Saviour."

A simple stone marks Annie's early grave,
No willow drops its tears above her head,
A few small tufts of heather sadly wave,
And two short words to tell where we have laid

"Annie Linn."

A slumb'ring babe awaits with her the hour,
When the loud trump shall wake the Moorland
Flower.

Some long, long years roll'd on, until at last
I saw my former life as some sweet dream
Which left a deep impression as it pass'd
That has survived, above the muddled stream
Of daily life; until one day I heard
A man lay ill with a disorder'd mind
And longed to see me, so without a word
I hastn'd out, his lone abode to find.
   It was a darksome prison-looking place,
Built for the poor unfortunate who live
And bear their Maker's likeness on their face,
But void of reason which alone can give
That stamp of man's superiority,
Or dress his actions in authority.

And there within a close dark cell was laid
All that was left of Aplin, pale and wan ;—
A miserable wreck, o'er which there play'd
The lamp's uncertain light, which shed upon
His wretched form a strange unearthly glare
That made me shudder ; as he turned his eyes

Upon me with a wild affrighted stare,
I pitied him I thought I should despise.
  He knew me, and a stifled groan escaped
His firm set lips, a sound of agony
Which seemed to rend in twain the gloom that drap'd
His heart in weighty folds of misery ;—
  He motion'd me unto him and I found
His gangrened limbs with iron fetters hung,
That trailed their loathsome length upon the ground
Or 'gainst the wall their links of sorrow swung.
  "Oh save me ! save me ! from this awful fate,
Rid me of life, or ease me of my pain,
And watch with me until as hours grow late
That torturing vision visits me again."
  "See where she comes ! there in that cold blue
      mist !
Ah you may see her now ; that wasted form
Fresh from the grave, round which the earthworms
      twist
And feed upon her beauty ;—lo ! and from
Those empty holes where once two bright blue eyes
Beam'd love, look now the loathsome reptiles creep ;—
And see her hair, matted with filth it lies
Upon her breast, where a dead child does sleep.
And see ! from off her arms the putrid flesh
Falls on the ground, and when it touches earth

Becomes a horrid imp, with torment fresh
From Hell it brings, from whence it took its birth,
And can you watch and see them all array'd
And quail not ? would it were a fantasy ;—
Nearer they come unless their flight be stayed
By some strong hand ;—and close behind them see
That changing form, 'tis beauteous Annie still,—
The cloud clears off, those hideous forms have gone,
Oh would they had not, for that look does chill
With horrors worse than they ;—but soft,—anon
It changes, and the smile which was her own
Usurps her face again, and seems to say
That she forgives me for the mighty wrong
I wrought her ;—now she quickly glides away.
    Oh, Annie, I have paid the penalty
A thousand times ;—in mercy pity me.
    " Ha ! ha ! ye fiends let loose again ye come
To vent your spleen,—Help ! help ! I feel them twine
Around my throat, I feel my coming doom
And cry for help who never pitied thine,
Oh save me, Annie, bid these torments fly,
Too late,—too late,— I faint,—I die,—I die."
    And sinking breathless on the floor he cast
His eyes on me, and raving breathed his last.
    A doleful story, friend, it sounds to you,
'Tis sorrowful,—'tis sad,—'tis strange,—but true.

Eyes beam as bright, smiles deck as dainty cheeks,   ·
As loving hearts beat anxiously within
As snowy breasts ; but fancy vainly seeks
Amongst them all, and finds no ANNIE LINN.—

CHARLES GOODALL, PRINTER, 16, WOODHOUSE LANE, LEEDS.

www.ingramcontent.com/pod-product-compliance
Lightning Source LLC
Chambersburg PA
CBHW022025190326
41519CB00010B/1606